Chemical Equilibrium

High School Chemistry: Chemical Equilibrium

Copyright © 2015 Vishal Mody

Published by Exam Masters Tutoring Service, a division of Ontario Inc.

17 Winnifred Avenue

Toronto, ON M4M 2X2

Acknowledgements

This Chemistry booklet is the culmination of years of tutoring Grade 11 and 12 high school chemistry students. Many thanks to all the students with whom I discussed the subject matter with for their valuable suggestions and corrections to the early versions. I am also indebted to the many students past and present that I have personally tutored through Exam Masters Tutoring Service, for challenging me to be the best I can be.

Authors Note

Many Grade 11 and 12 students have a lot of trouble with concepts in Chemistry. These students often are not taught specific concepts well and without that solid foundation, it becomes very hard to understand the more difficult concepts.

This chemistry booklet was created to help students specifically with the topic of Chemical Equilibrium in chemistry. This booklet has been made extremely concise yet explains the concepts in detail at the same time. Remember, that this booklet is not designed to be your main study source, but rather, as an adjunct to your school teacher's notes. There are also lots of practice questions with detailed solutions at the end to solidify the concepts you have learned.

Best wishes in your studies,

Exam Masters[TS]

Table of Contents

1

Chemical Equilibrium:

Chemical Equilibrium

If a reversible reaction is allowed to continue for a considerable long time, without changing the conditions, there is no further change in composition of the reaction mixture this means that the reaction has achieved a state of chemical equilibrium. Once this state is attained, it will last forever if undisturbed.

For example, let us consider a reaction in which A reacts with B to produce C and D.

$$A\ (g)\ +\ B\ (g)\ \rightleftharpoons\ C\ (g)\ +\ D\ (g)$$

Suppose that all the substances are in gaseous state. Let us consider a graph between time and concentrations for reactants and products.

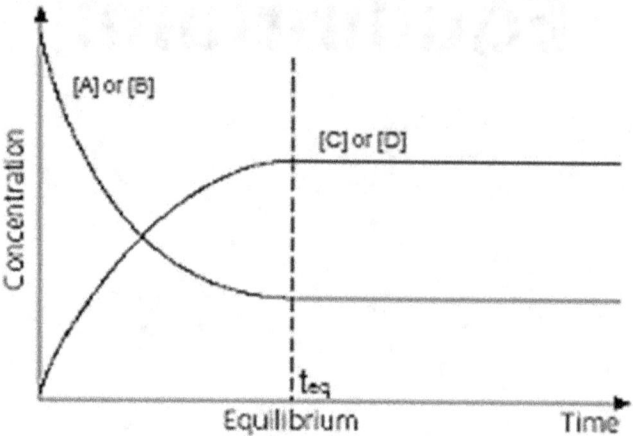

Figure-1: Graph between time and concentration

Let the initial concentrations A and B be equal. As the time goes on, concentration of A and B decrease, at first quite rapidly but later slowly. Eventually, the concentrations of A and B level off and become constant. The initial concentrations of C and D are zero. As the time

The Le-Chatelier's Principle:_ _ _ _ _ _ _ _ _ _ _ _ _ _ _

passes the product C and D are formed. Their concentrations increase rapidly at first and then level off. At the time of equilibrium, the concentrations become constant. In this way chemical equilibrium is attained and state of equilibrium is reached.

Let us consider a reversible reaction between hydrogen gas and iodine vapors to form hydrogen iodide at 425°C. At equilibrium three components will be present in definite proportions in the reaction mixture as shown below:

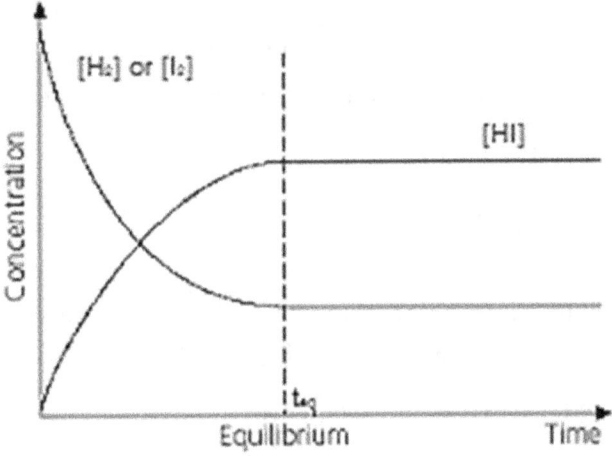

Figure-2: Graph between time and concentration

The equilibrium is established when the rising curve of product HI and the falling curve of reactants [H2] AND [I2] become parallel to time axis

$$H_2 \text{ (g)} + I_2 \text{ (g)} \rightleftharpoons 2 \text{ HI (g)}$$

Chemical Equilibrium

The same equilibrium mixture is obtained irrespective whether the reaction starts by mixing hydrogen and iodine or by decomposition of hydrogen iodide. The situation suggests two possibilities of the state of reaction at equilibrium.

1. All reactions cease at equilibrium so that the system becomes stationary.

2. The forward and reverse reactions are taking place simultaneously at exactly the same rate.

Now it is unanimously accepted that the second condition prevail in a reversible reaction at equilibrium. It is known as state of dynamic equilibrium.

Law of Mass Action:

A state of dynamic equilibrium helps to determine the composition of reacting substances and the products at the equilibrium. The law of mass action states that:

"The rate at which the reaction proceeds is directly proportional to the product of the active masses of the reactants".

The term active mass represents the concentration in mold m-3 of the reactants and products for a dilute solution.

The Le-Chatelier's Principle: _ _ _ _ _ _ _ _ _ _ _ _ _ _

QUESTIONS

1. Define the term "state of equilibrium".

2. When a graph is plotted between time on x-axis and concentration on y-axis, the curve becomes parallel to time axis at certain stage. Why?

3. What are the two possibilities of state of reaction at equilibrium? And which one of these is most acceptable?

4. State the law of mass action and what is meant by active mass of a substance?

2

The Le-Chatelier's Principle:

Chemical Equilibrium

Le-Chatelier studied the effects of concentrations, pressure and temperature on equilibrium. According to this principle:

"If stress is applied to a system at equilibrium, the system acts in such a way so as to nullify, as far as possible, the effects of that stress"

The system cannot completely cancel the effect of change, but will minimize it. The le-chatelier's principle has wide range of applications in physical and chemical equilibrium.

A. **Effect Of Change In Concentration:**

In order to understand the effect of change in concentration on the reversible reaction, consider the reaction in which BiCl2 reacts with water to give a white insoluble compound BiOCl.

$$BiCl_3 + H_2O \quad BiOCl + 2HCl$$

The equilibrium constant expression for above reaction can be written as:

$$KC=\left[\frac{BiOCl][HCl]^2}{[BiCl_3][H_2O]} \right.$$

Aqueous solution of BiCl3 is cloudy, because of hydrolysis and formation of BiOCl. If a small amount of HCl is added to this solution, it will disturb the equilibrium and force the system to move in such a way so the effect of addition of HCl is minimized. The reaction will move in the backward direction to restore the equilibrium again and a clear solution will be obtained. However, if the water is added to the above solution the system will move in the forward direction and the solution will again become cloudy. The shifting of reaction to forward and backward direction by disturbing the concentration is just according to Le-Chatelier's principle.

So, in general, we can say that adding of a substance among the reactants or the removal of a substance among the products at equilibrium stage disturbs the equilibrium position and the reaction is shifted to forward direction. Similarly, the addition of a substance among the products or the removal of a substance among the reactants will shift the equilibrium towards the backward direction. Removing one of the products formed can therefore increase the yield of reversible reaction. The value of KC however remains constant.

B. **Effect Of Change In Pressure Or Volume:**

The change in pressure or volume are important only for the reversible gaseous reactions where the number of moles of reactants and products are not equal. Le-Chatelier's principle plays an important role, to predict the position and direction of the reaction. For example, consider formation of SO3 gas from SO2 gas and O2 gas.

$$2SO_2(g) \ + \ O2(g) \ \rightleftharpoons \ 2SO_3(g)$$

This gas phase reaction proceeds with the decrease in the number of moles and hence decrease in volume at equilibrium stage. When the reaction reaches the equilibrium stage, the volume of the equilibrium mixture is less than the volume of reactants taken initially. If one decrease the volume further at equilibrium stage, the reaction is disturbed. It will move to forward direction to minimize the effect of disturbance. It establishes a new equilibrium position while KC remains constant. The reverse happens when the volume is increased or pressure is decreased at equilibrium stage.

Those gaseous reactions in which number of moles of reactants and products are same, are not affected by change in pressure or volume. Same is the case for reactions in which the participating substances are either liquids or solids.

C. Effect Of Change In Temperature:

Most of the reversible chemical reactions are disturbed by change in temperature. If we consider heat as component of equilibrium system, a rise in temperature adds heat to the system and a drop in temperature removes heat from the system. According to Le-Chatelier's principle, if we increase temperature, it favors the endothermic reactions and a temperature decrease favors the exothermic reactions.

The equilibrium constant changes by change of temperature, because the equilibrium position shifts without any substance being added or removed. Consider the following exothermic reaction in gas phase at equilibrium taking place at a known temperature.

$$CO\ (g)\ +\ H_2O\ (g)\ \rightleftharpoons\ CO_2\ (g)\ +\ H_2\ (g)\qquad \Delta H = -41.84 kJ\ mol^{-1}$$

At equilibrium stage, if we take out heat and keep the system at this new lower temperature, the system will readjust itself, so as to compensate the loss of heat energy. Thus, more of CO and H_2O

molecules will react to form CO_2 and H_2 molecules, thereby, releasing heat because the reaction is exothermic in the forward direction. It means by decreasing the temperature, we shift the initial equilibrium position to the right until a new equilibrium position is established. On the contrary, heating the reaction at equilibrium will shift the reaction to the backward direction because the backward reaction is endothermic.

D.   **Change In Solubility:**

An interesting feature of Le-Chatelier's principle is the effect of temperature on solubility. Consider a salt such as KI. It dissolves in water and absorbs heat.

$$KI \rightleftharpoons (s) \ KI \ (aq) \qquad \Delta H = 21.4Kj \ mole^{-1}$$

Let us consider a saturated solution of KI in water at a given temperature. It has reached equilibrium at this temperature. A rise in temperature at equilibrium favors more dissolution of salt. Equilibrium is shifted to the forward direction. On the other hand, cooling will favor crystallization of salt. Hence the solubility of KI in water must increase with increase in temperature.

For some salts the heat of solution is close to zero i.e. heat is neither evolved nor absorbed. The solubility of these salts in water is not affected by the change in temperature. Formation of aqueous solution of NaCl is an example of such a salt.

Those substances, whose heats of solution are negative (exothermic, decrease their solubilities by increasing temperature, as LiCl and Li_2CO_3 etc.

E. **Effect Of Catalyst On Equilibrium Constant:**

In most of the reversible reactions the equilibrium is not always reached with in a suitable short time. So, appropriate catalyst is added. A catalyst does not affect the equilibrium position of the reaction. It increases the rates of both forward and backward reactions and this reduces the time to attain the state of equilibrium. Actually a catalyst lowers the energy of activation of both forward and reverse steps by forming a new path of the reaction.

QUESTIONS

5. The aqueous solution of $BiCl_3$ is cloudy. State why it clears on addition of HCl and why it again turns cloudy on addition of water to solution?

6. How can we increase the yield of a reversible reaction by changing the concentration of reactants or products?

7. how the yield of SO3 is increased by decreasing volume of equilibrium mixture?

$$2SO_2\,(g) \ + \ O_2\,(g) \ \rightleftharpoons \ 2SO_3\,(g)$$

8. What type of reversible reactions are not effected by change in volume and pressure of equilibrium constant?

The Le-Chatelier's Principle: _ _ _ _ _ _ _ _ _ _ _ _ _ _

9. Consider the following reaction in gas phase at equilibrium taking place at a known temperature.

    $$CO\ (g)\ +\ H_2O\ (g) \rightleftharpoons CO_2\ (g)\ +\ H_2\ (g) \qquad ?H = -41.84kJ\ mol^{-1}$$

    In which direction the reaction will proceed if we lower the temperature of reaction mixture?

10. what are the different types of salts on basis of heats of solution? Also give at least one example of each.

11. How a catalyst increases the rate of reaction?

3

Forward and Reverse Reaction Rates and Equilibrium

Chemical Equilibrium

A chemical reaction can take place in both directions, i.e. forward and reverse, but in some cases the tendency of reverse reaction is very small and is negligible. For example, sodium reacts with water to form sodium hydroxide and hydrogen gas. That's a forward reaction.

$$2Na \ (s) \ + \ 2H_2O \ (l) \ \rightarrow \ 2NaOH \ (aq) \ +H_2 \ (g)$$

The tendency for hydrogen to react with sodium hydroxide to form sodium and water i.e. reverse reactions is negligible at normal temperature. This is an irreversible reaction.

Let us consider another example of the reaction between two parts of hydrogen and one part of oxygen in the presence of electric spark at normal temperature and pressure. The reaction occurs stoichiometrically according to the following chemical equation:

$$2H_2 \ (g) \ + \ O_2 \ (g) \ \rightarrow \ 2H_2O \ (l)$$

If hydrogen and oxygen are present in correct proportion, there will be no residual gases i.e. hydrogen and oxygen. If the product is heated to a temperature of 1500 °C, a noticeable quantity of H_2O decomposes, producing hydrogen and oxygen. It means that reverse reaction does occur, but only at higher temperature. It is very likely that the reverse reaction occurs at low temperature, but it is too small to be noticeable. The reaction between stoichiometric amounts of hydrogen and oxygen proceeds to completion in the presence of electric spark. Such reactions are called irreversible reactions and they take place in one direction only.

Now, consider a reaction between nitrogen and hydrogen at 450 °C under high pressure in the presence of iron as a catalyst.

$$Fe/450C$$

$$N_2 (g) + 3H_2 (g) \quad \overset{\text{High}}{\underset{\text{pressure}}{\rightleftharpoons}} \quad 2NH_3 (g)$$

The reaction mixture, after some time will contain all the three species, i.e nitrogen, hydrogen and ammonia. No matter how long the reaction is allowed to continue, the percentage composition of the species present remain constant. The conditions are favorable for the forward as well as for a reverse reaction to occur to a measureable extent. This type of reaction is described as a reversible reaction.

Initially, in the start of the reaction the rate of forward reaction is very fast. With time the rate decreases as the reactants are converted into the products and then the reaction achieves an equilibrium state. If a reversible reaction is continue to allow for a considerable period of time without changing the conditions, there will be no further change in composition of the reaction mixture. This is the stage of chemical equilibrium. Equilibrium is of two types:

1. **Static Equilibrium:**

It occurs when all particles in the reaction are at rest and there is no motion between reactants and products. For example, graphite turning into diamond. This reaction is considered at static equilibrium after it occurs because there are no forces acting on the reactant (graphite) and products (diamond).

Another example is the burning of paper. Once paper is burned the reaction achieves a static equilibrium

2.   **Dynamic equilibrium:**

At dynamic equilibrium, reactants are converted to products and products are converted to reactants at an equal and constant rate. Reactions do not necessarily and most often do not end up with equal concentrations. Equilibrium is the state of equal, opposite rates, not equal concentrations. At dynamic equilibrium the rates of forward and reverse reactions are equal. So at equilibrium:

$$RATE_{forward} \quad = \quad RATE_{reverse}$$

The forward and reverse reaction rates can be changed by changing the concentration of reactants or products, by change in pressure or volume, by temperature change or by catalyst according to Le-Chatelier's principle.

QUESTIONS

12. What are irreversible reaction? Also give an example.

13. What are reversible reactions? Give an example of a reversible reaction.

4

Impact on Society: The Haber-Bosch Process

Chemical Equilibrium

The process of ammonia synthesis was developed by German chemist F.Haber and was first used in 1933. This process is a good example of applying equilibrium principle in the industrial process. The chemical equation for this reaction is as under:

$$N_2 (g) + \rightleftharpoons 3H_2 (g) \quad 2NH_3 (g) \qquad \Delta H = -92.46 \text{ kJ}$$

According to this equation and Le-Chatelier's principle, we can conclude that there are three ways to maximize the yield of ammonia:

a. By continual withdrawal of ammonia after intervals, the equilibrium will shift to forward direction.

b. Increase the pressure to decrease the volume of the reaction vessel. Four moles of the reactants combine to give two moles of the products. High pressure will shift the equilibrium position to right to give more and more ammonia.

c. Decreasing the temperature will shift it to the forward direction according to Le-Chatelier's principle.

So high pressure, low temperature and continual removal of ammonia will give the maximum yield of ammonia. Following graph will show the maximum yield of ammonia.

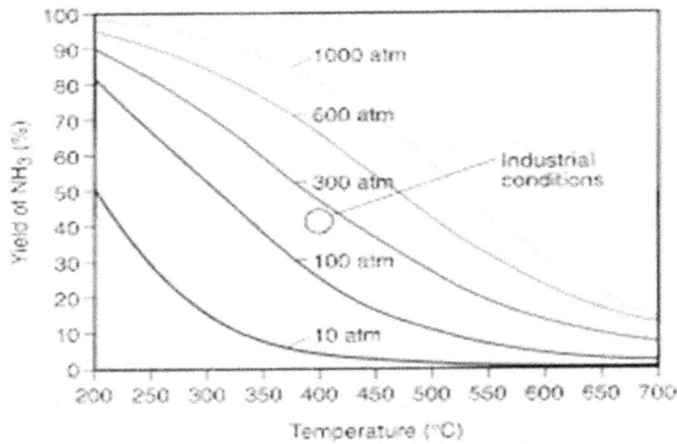

At very high pressure and low temperature, the yield of ammonia is high but the rate of formation is low. Industrial conditions showed by circle are between 200 and 300 atmospheres at about 400 °C.

Although the yield of ammonia is favored at low temperature, but the rate of its formation does not remain favorable. The rate becomes so slow and the process is rendered uneconomical. We need a compromise to optimize the yield and the rate. The temperature is raised to a moderate level and the catalyst is used to increase the rate. If we want to achieve same rate without a catalyst, then it will require a much higher temperature which will in turn lowers the yield. Hence the optimum conditions are the pressure of 200 to 300 atm and temperature around 400 °C. The catalyst is the pieces of iron crystals embedded in fused mixture of MgO, Al_2O_3 and SiO_2.

The equilibrium mixture has 35% by volume of ammonia. The mixture is cooled by refrigeration coils until ammonia condenses and is removed. Since, boiling points of nitrogen and hydrogen are very low, they remain in the gaseous state and are recycled by pumps back into the reaction chamber.

Impact on the Society:

The Haber process has a wide range of impacts on society, these include:

1. Nearly 13% of all nitrogen fixation on earth is accomplished industrially through Haber's process. This process synthesizes approximately 110 million tons of ammonia in the world. About 80% of this is used for the production of fertilizers. These fertilizers have quadrupled the productivity of agricultural land.

2. Due to increase production of food, the human population has increased a lot.

3. Ammonia can also be used for the production of nylon and other polymers.

Negative effects:

1. It can cause serious imbalance in the nitrogen cycle.

2. This process consumes a lot of energy.

3. It can cause many negative effects on soil organisms and soil organic matter.

4. Excess runoff can cause ocean dead zones.

5. It can be used for the manufacture of explosives.

QUESTIONS

14. How the yield of ammonia can be increased in Haber-Bosch process by applying Le-Chatelier's principle?

15. What are the optimum conditions needed for getting maximum yield of ammonia at industrial scale?

16. Why the ammonia is not produced at very high pressure and very low temperature at industrial scale, while its yield is also maximum at these extreme conditions?

17. Normally the reaction rate of ammonia formation reaction increases with increase in temperature but it greatly reduces yield of ammonia. Is there any method to increase reaction rate while keeping higher yield?

18. what is the importance of haber-bosch process in society?

5

Equilibrium Constant Expressions

Chemical Equilibrium

Equilibrium constant is the ratio of the products of the concentrations of the products to the product of concentrations of the reactants.

According to law of mass action, the rate at which the reaction proceeds is directly proportional to the product of the active masses of the reactants.

Now, consider a general reaction in which A and B are reactants and C and D are the products. The reaction can be represented as:

$$A + B \rightleftharpoons C + D$$

The equilibrium concentrations of A, B, C and D can be represented in square brackets as [A], [B], [C] and [D] and their unit will be moles dm-3. According to law of mass action, the rate of forward reaction, is proportional to the product of molar concentrations of A and B.

Rate of forward reaction (R_f) ∞ [A] [B]

Or

$$R_f = k_f [A] [B]$$

Where kf is the proportionality constant and is called rate constant for forward reaction and Rf is the rate of forward reaction. Similarly, the rate of reverse reaction (Rr) is given by:

Rate of reverse reaction (R_r) ∞ [C] [D]

Or

Equilibrium Constant Expressions

$$R_r = k_r [C] [D]$$

Where kr is the proportionality constant and is called the rate constant for backward reaction. Here C and D are reactants for backward step.

At equilibrium,

$$R_f = R_r$$

Or
$$kf [A] [B] = K_r [C] [D]$$

On rearranging we get:

$$\left(\frac{k_f}{k_r}\right) = \frac{[C][D]}{[A][B]}$$

Let

$$\left(\frac{k_f}{k_r}\right) = (k_c)$$

So,

$$k_c = \frac{[C][D]}{[A][B]}$$

Chemical Equilibrium

The constant kc is called the equilibrium constant of the reaction. Kc is the ratio of two rate constants. Conventionally, while writing equilibrium constant, the products are written as numerator and the reactants as denominator.

$$k_c = \frac{[products]}{[reactants]}$$

Or

$$k_c = \frac{rate\ constant\ for\ forward\ step}{rate\ constant\ for\ reverse\ step}$$

For a more general reaction,

$$a\,A \ + \ b\,B \ \rightleftharpoons \ c\,C \ + \ d\,D$$

Where a, b, c and d are the coefficients of balanced chemical equation. They are number of moles of A, B, C and D respectively in the balanced equation.

The equilibrium constant is given by:

$$k_c = \frac{[C]^c[D]^d}{[A]^a[B]^b}$$

Hence the coefficients in the equation appear as exponents of the terms of concentrations in the equilibrium constant expression.

Equilibrium Constant Expressions

Units of Equilibrium Constants:

If the reaction has equal number of moles on the reactant and product sides, then equilibrium constants has no units. When the number of moles is unequal then it has units related to the concentration or pressure. But it is a usual practice to not to write the units with kc values.

Applications of Equilibrium Constant:

The value of equilibrium constant is specific and remains constant at a particular temperature. It can give following information.

1. *Direction Of Reaction:*

    We know that,

    $$k_c = \frac{[products]}{[reactants]}$$

    The direction of a chemical reaction at any particular time can be predicted by means of [products]/[reactants] ratio, calculated before the reaction attains equilibrium. The value of [product]/[reactant] ratio leads to one of the following three possibilities.

    a. The ratio is less than kc. This implies that more of the product is required to attain the equilibrium, therefore, the reaction will proceed in the forward direction.

b. The ratio is greater than kc. It means that the reverse reaction will occur to attain the equilibrium.

c. When the ratio is equal to kc then the reaction is at equilibrium.

2. *Extent Of Reaction:*

a. If the equilibrium constant is very large, this indicates that the reaction is almost complete.

b. If the value of kc is small, it reflects that the reaction does not proceed appreciably in the forward direction.

c. If the value of kc is very small, this shows a very little forward reaction.

3. *The Effect Of Conditions On The Position Of Equilibrium:*

Equilibrium constant and position of equilibrium are two different entities. Kc is equilibrium constant and has a constant value at a particular temperature whereas the ratio of products to reactants in equilibrium mixture is described as the position of equilibrium and it can change if the external conditions e.g. temperature, pressure and concentrations are altered. If k_c is large the position of equilibrium lies on the right and if it is small, the position of the equilibrium lies on the left, for a reversible reaction.

Equilibrium Constant Expressions

QUESTIONS

19. How the value of equilibrium constant effects the direction of reaction?

20. How the equilibrium constant relates to position of equilibrium?

ANSWERS

Answer-1:   If a reversible reaction is allowed to continue for a considerable long time, without changing the conditions, there is no further change in composition of the reaction mixture this means that the reaction has achieved a state of chemical equilibrium. Once this state is attained, it will last forever if undisturbed.

Answer-2:

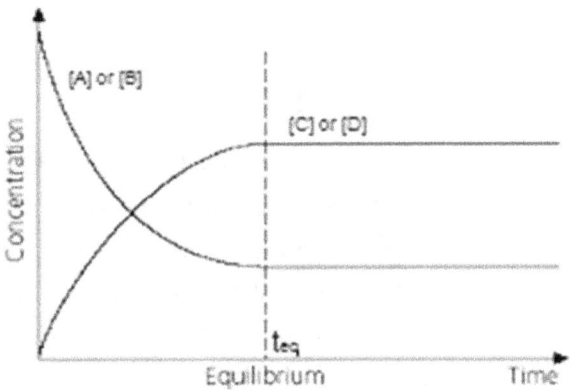

Dissolve a pure sample of the compound in freshly distilled water. If the concentration of hydrogen ions goes up, the compound is an acid. There are several ways to detect the rise in hydrogen ion concentration. Certain dyes called indicators change color when exposed to acid solutions; for example, blue litmus turns pink. You can also use a pH meter; a pH of lower than 7 corresponds to an acidic solution. (pH goes down as hydrogen ion concentration goes up...)

Answer-3: There are two possibilities of the state of reaction at equilibrium.

3. Static Equilibrium:

It occurs when all particles in the reaction are at rest and there is no motion between reactants and products. For example, graphite turning into diamond. This reaction is considered at static equilibrium after it occurs because there are no forces acting on the reactant (graphite) and products (diamond).

Another example is the burning of paper . Once paper is burned the reaction achieves a static equilibrium

4. **Dynamic equilibrium:**

At dynamic equilibrium, reactants are converted to products and products are converted to reactants at an equal and constant rate. Reactions do not necessarily and most often do not end up with equal concentrations. Equilibrium is the state of equal, opposite rates, not equal concentrations. At dynamic equilibrium the rates of forward and reverse reactions are equal. So at equilibrium:

$$RATE_{forward} = RATE_{reverse}$$

Now it is unanimously accepted that the second condition prevail in a reversible reaction at equilibrium. It is known as state of dynamic equilibrium.

Answer-4:   The law of mass action states that:

"The rate at which the reaction proceeds is directly proportional to the product of the active masses of the reactants".

The term active mass represents the concentration in mold m-3 of the reactants and products for a dilute solution.

Chemical Equilibrium

Answer-5: BiCl3 + H2O BiOCl + 2HCl

The equilibrium constant expression for above reaction can be written as:

$$K_C = \frac{[BiOCl][HCl]^2}{[BiCl_3][H_2O]}$$

Aqueous solution of BiCl3 is cloudy, because of hydrolysis and formation of BiOCl. If a small amount of HCl is added to this solution, it will disturb the equilibrium and force the system to move in such a way so tha effect of addition of HCl is minimized. The reaction will move in the backward direaction to restore the equilibrium again and a clear solution will be obtained. However, if the water is added to the above solution the system will move in the forward direction and the solution will again become cloudy. The shifting of reaction to forward and backward direction by disturbing the concentration is just according to Le-Chatelier's principle.

ANSWERS

Answer-6: According to Le-Chatelier's principle adding of a substance among the reactants or the removal of a substance among the products at equilibrium stage disturbs the equilibrium position and the reaction is shifted to forward direction. Similarly, the addition of a substance among the products or the removal of a substance among the reactants will shift the equilibrium towards the backward direction. Removing one of the products formed can therefore increase the yield of reversible reaction. The value of K_C however remains constant.

Answer-7: This gas phase reaction proceeds with the decrease in the number of moles and hence decrease in volume at equilibrium stage. When the reaction reaches the equilibrium stage, the volume of the equilibrium mixture is less than the volume of reactants taken initially. If one decrease the volume further at equilibrium stage, the reaction is disturbed. It will move to forward direction to minimize the effect of disturbance. Hence, we can get higher yield of SO_3.

Answer-8: Those gaseous reactions in which number of moles of reactants and products are same, are not affected by change in pressure or volume. Same is the case for reactions in which the participating substances are either liquids or solids

Answer-9:   According to Le-Chatelier's principle, if we increase temperature, it favors the endothermic reactions and a temperature decrease favors the exothermic reactions. As this is an exothermic reaction so at equilibrium stage, if we take out heat and keep the system at this new lower temperature, the system will readjust itself, so as to compensate the loss of heat energy. Thus, more of CO and H_2O molecules will react to form CO_2 and H_2 molecules, thereby, releasing heat because the reaction is exothermic in the forward direction.

Answer-10:   Salts having heats of solution that has positive value (endothermic, increase their solubilities by increasing temperature. Example is aqueous solution of KI.

Salts having heat of solution close to zero i.e. heat is neither evolved nor absorbed. The solubility of these salts in water is not affected by the change in temperature. Formation of aqueous solution of NaCl is an example of such a salt.

Salts having heats of solution that has negative value (exothermic, decrease their solubilities by increasing temperature, as LiCl and Li_2CO_3 etc.

ANSWERS

Answer-11: In most of the reversible reactions the equilibrium is not always reached with in a suitable short time. So, appropriate catalyst is added. A catalyst does not affect the equilibrium position of the reaction. It increases the rates of both forward and backward reactions and this reduces the time to attain the state of equilibrium. Actually a catalyst lowers the energy of activation of both forward and reverse steps by forming a new path of the reaction.

Answer-12: The reactions are called irreversible reactions if they take place in one direction only.

For example, sodium reacts with water to form sodium hydroxide and hydrogen gas. That's a forward reaction.

$$2Na\ (s)\ +\ 2H_2O\ (l)\ \rightarrow\ 2NaOH\ (aq)\ +H_2\ (g)$$

The tendency for hydrogen to react with sodium hydroxide to form sodium and water i.e reverse reactions is negligible at normal temperature. This is an irreversible reaction

Answer-13:   The reactions are called reversible reactions if they take place in both forward and reverse directions.

For example, a reaction between nitrogen and hydrogen at 450 °C under high pressure in the presence of iron as a catalyst.

$$N_2 (g) + 3H_2 (g) \underset{\text{High pressure}}{\overset{\text{Fe/450C}}{\rightleftarrows}} 2NH_3 (g)$$

The reaction mixture, after some time will contain all the three species, i.e nitrogen, hydrogen and ammonia. No matter how long the reaction is allowed to continue, the percentage composition of the species present remain constant. The conditions are favorable for the forward as well as for a reverse reaction to occur to a measureable extent.

ANSWERS

Answer-14: $N_2 (g) + 3H_2 (g) \rightarrow 2NH_3 (g)$
$\Delta H = -92.46 \text{ kJ}$

According to this equation and Le-Chatelier principle we can conclude that there are three ways to maximize the yield of ammonia:

- By continual withdrawal of ammonia after intervals, the equilibrium will shift to forward direction.

- Increase the pressure to decrease the volume of the reaction vessel. Four moles of the reactants combine to give two moles of the products. High pressure will shift the equilibrium position to right to give more and more ammonia.

- Decreasing the temperature will shift it to the forward direction according to Le-Chatelier's principle.

So high pressure, low temperature and continual removal of ammonia will give the maximum yield of ammonia.

Answer-15:   The optimum conditions are the pressure of 200 to 300 atm and temperature around 400 °C. The catalyst is the pieces of iron crystals embedded in fused mixture of MgO, Al_2O_3 and SiO_2.

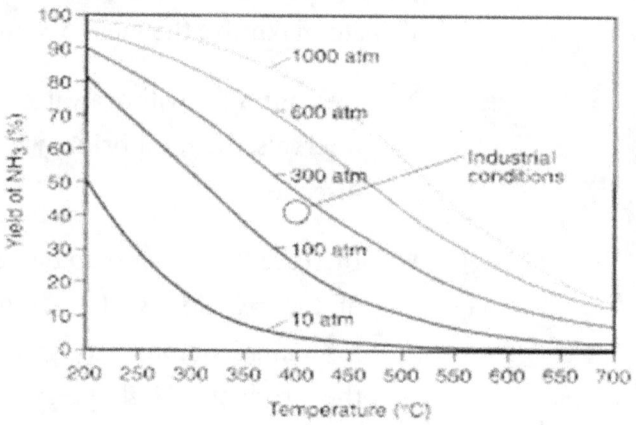

Answer-16:   At very high pressure and low temperature, the yield of ammonia is high but the rate of formation is low. Industrial conditions are between 200 and 300 atmospheres pressure at about 400 °C temperature. Although the yield of ammonia is favored at low temperature, but the rate of its formation does not remain favorable. The rate becomes so slow and the process is rendered uneconomical. We need a compromise to optimize the yield and the rate.

ANSWERS

Answer-17: The temperature is raised to a moderate level and the catalyst is used to increase the rate. If we want to achieve same rate without a catalyst, then it will require a much higher temperature which will in turn lowers the yield. Hence the optimum conditions are the pressure of 200 to 300 atm and temperature around 400 °C. The catalyst is the pieces of iron crystals embedded in fused mixture of MgO, Al_2O_3 and SiO_2.

Answer-18: The temperature is raised to a moderate level and the catalyst is used to increase the rate. If we want to achieve same rate without a catalyst, then it will require a much higher temperature which will in turn lowers the yield. Hence the optimum conditions are the pressure of 200 to 300 atm and temperature around 400 °C. The catalyst is the pieces of iron crystals embedded in fused mixture of MgO, Al_2O_3 and SiO_2.

Answer-19:   Direction of Reaction:

We know that,

$$k_c = \frac{[products]}{[reactants]}$$

The direction of a chemical reaction at any particular time can be predicted by means of [products]/[reactants] ratio, calculated before the reaction attains equilibrium. The value of [product]/[reactant] ratio leads to one of the following three possibilities.

- The ratio is less than k_c. This implies that more of the product is required to attain the equilibrium, therefore, the reaction will proceed in the forward direction.

- The ratio is greater than k_c. It means that the reverse reaction will occur to attain the equilibrium.

- When the ratio is equal to k_c then the reaction is at equilibrium.

ANSWERS

Answer-20: Equilibrium constant and position of equilibrium are two different entities. K_c is equilibrium constant and has a constant value at a particular temperature whereas the ratio of products to reactants in equilibrium mixture is described as the position of equilibrium and it can change if the external conditions e.g. temperature, pressure and concentrations are altered. If k_c is large the position of equilibrium lies on the right and if it is small, the position of the equilibrium lies on the left, for a reversible reaction.